INTERNATIONAL CENTRE FOR MECHANICAL SCIENCES

COURSES AND LECTURES - No. 37

TATIANA V. BAZHENOVA

ACADEMY OF SCIENCES, MOSCOW

SHOCK WAVES IN REAL GASES

COURSE HELD AT THE DEPARTMENT
OF HYDRO-AND GAS-DYNAMICS
JULY 1970

UDINE 1973

SPRINGER-VERLAG WIEN GMBH

Originally published by Springer-Verlag Wien-New York in 1972

ISBN 978-3-211-81219-8   ISBN 978-3-7091-2854-1 (eBook)

DOI 10.1007/978-3-7091-2854-1

P R E F A C E

The problem is considered of the effect of real properties of gases on the reflection of shock waves. The results are presented in the experimental study of unsteady propagation of a reflected shock wave in a reacting gas and Mach reflection in a gas with excited degrees of freedom. Numerical methods of calculation of Mach reflection in a real gas are described and comparison with the experiment is presented.

Udine, July 1970

T.V. BAZHENOVA

1. NUMERICAL METHODS OF CALCULATION OF REFLECTION OF SHOCK WAVES IN REAL GASES

In considering the properties of strong shock waves consideration should always be given to the real properties of the gas, since the temperature increase taking place behind a shock wave produces physicochemical transformations, i.e., excitation of internal degrees of freedom and ionization, in the flow behind the wave.

The relationship between the composition of a gas and its variables (temperature and pressure) on equilibrium is determined by known thermodynamic laws. Equilibrium in a reacting medium ensues when the rates of disintegration processes (dissociation and ionization) are equal to the rate of recombination processes which take place simultaneously. Detailed tables of the composition and parameters of air, carbon dioxide and nitrogen behind shock waves have been calculated for these conditions in a wide range of pressures at temperatures of up to 20,000°K. The state of equilibrium is established behind a shock wave at a finite distance from the front, for which reason a certain time, known as the relaxation time, is needed for the equilibrium to come into effect. The flow behind a shock wave in the nonequilibrium region has different properties, i.e., the degree of dissociation and ionization increases in this region

from zero across the front to the equilibrium value, and this
is accompanied by a several-thousand degree drop in the gas
temperature, since the kinetic energy of the gas molecules is
used up for exciting internal degrees of freedom. The fact that
the gas behind a shock wave is not in equilibrium affects the
shape of the shock wave which is produced at bodies in supersonic flow, the distance between the front and the body, the
interaction between shock waves and the structure of the supersonic flow.

The rate at which equilibrium values of variables
behind a shock wave are established can be calculated if the
rate constants of all the elementary processes taking place in
the gas are known. However, these constants can at present be
calculated only for individual reactions, at the time when heating of each gas is accompanied by tens of different reactions.
Consequently, experimental studies must be made of the formation of shock waves and the properties of flow in the nonequilibrium region behind shock wave must be determined in order to
obtain information on the rates of physicochemical transformations.

As it will further be seen, real gases proper-
ties in some cases lead not only to the quantitative, but also
to the qualitative peculiarities of shock waves behaviour, for
instance, in the case of reflection of normal and oblique shock
waves.

Let us consider the effect of real gases properties on the calculation of the parameters of a reflected shock wave. At first, we shall consider simple cases, and then more complicated ones.

The state behind the plane stationary shock wave can be determined by resorting to conservation laws written in the coordinate system moving with the shock wave ;

1) equation of conservation of mass :

$$\rho_1 \upsilon_1 = \rho_0 \upsilon_0 = G \; ,$$

2) equation of conservation of momentum :

$$\rho_1 \upsilon_1^2 + p_1 = \rho_0 \upsilon_0^2 + p_0 = I \; ,$$

3) equation of conservation of energy :

$$\upsilon_1^2/2 + h = \upsilon_0^2/2 + h_0 = H \; ,$$

4) equation of state

$$p_1 m_1 = \rho_1 R T_1 \; .$$

Here the following denotions are used :

ρ – density, υ –velocity, p –pressure,

h – enthalpy, m – molecular weight.

Symbol (o) stands for the initial state, symbol (I) stands for the state behind the shock wave moving at the constant velocity υ_0 .

If the gas is ideal, the connection between the values h and T is given by the equation

$$h = c_p T = \frac{\gamma}{\gamma - 1} \frac{P}{\rho} \; ;$$

wherein :

c_p is the heat capacity under the constant pressure, and γ is the heat capacity ratio $\gamma = c_p / c_\sigma$.

If the gas is real, the molecular weight can be changed behind the shock wave. The mean molecular weight and the enthalpy are the functions of the concentration of the gas components x_i

$$m = \sum m_i x_i \qquad h = \sum x_i h_i \; .$$

The composition of the gas depends upon the temperature, because the value of the equilibrium constant K_p depends on the values of the statistical sums Q_i . The values of Q_i as a function of the temperature T are given in the tables. The functions $Q_i(T)$ cannot be written in the analytical form, as they depend on the spectroscopical constants w and B which are obtained experimentally only. Because of this we must use numerical methods for calculating the parameters of shock waves in a real gas. The system of conservation laws can be transformed to the following form :

$$v_1 = \frac{I}{2G} \pm \sqrt{\frac{I^2}{2G^2} - \frac{RT_1}{m_1}}$$

$$h + \frac{I}{4G^2} \sqrt{I^2 - \frac{4G^2 RT_1}{m_1}} - \frac{RT_1}{2m} = \lambda - \frac{I^2}{4G^2}$$

This system can be solved by the trail-and-error method, if we take into account the fact that the excitation of internal degrees of freedom of molecules practically do not affect the value of the pressure behind the shock wave P_1

This method is very convenient if we must calculate the partially frozen state behind a shock wave, where only some degrees of freedom are excited. In this case we take the value of H ,which is calculated in the assumption of a partially frozen state, for example :

(1) Internal degrees of freedom are excited but no dissociation takes place.

(2) Some degrees of freedom are excited and one or two of them are frozen.

The state of the gas behind the reflected wave (2) and the velocity of the reflected wave U_2 could be determined from conservation laws written in a coordinate system moving with the reflected wave :

$$\rho_1 (v_1 + u_2) = \rho_2 u_2 \qquad \frac{(v_1 + u_2)^2}{2} + h_1 = \frac{u_2^2}{2} + h_2$$

$$\rho_1 (v_1 + u_2)^2 + p_1 = \rho_2 u_2^2 + p_2 \qquad p_2 = \frac{\rho_2 R T_2}{m_2}$$

This system can be reduced to the following form :

$$u_2 = - \frac{v_1^2 + \dfrac{R T_1}{m_1} - \dfrac{R T_2}{m_2}}{2 \, v_1} + \sqrt{\left[\frac{v_1^2 + \dfrac{R T_1}{m_1} - \dfrac{R T_2}{m_2}}{2 \, v_1} \right]^2 + \frac{R T_2}{m_2}}$$

$$v_1 u_2 + \frac{v_1^2}{2} + h_1 = h_2 \left(T_2 \, \varphi_2 \right) .$$

The calculation is made by the trial–and–error method, taking into account the proper values of H_1 and H_2 . The inclination angle of the attached shock wave φ in a real gas could be calculated by the shock polar method, if the wedge angle θ and the gas properties are known. The tangent component of the velocity of the gas behind the oblique shock wave is equal to that in front of the shock :

$$v_{1t} = v_{0t} .$$

Then, the relation between the normal components of velocity is equal to the relation between the densities :

$$\frac{v_{1n}}{v_{on}} = \frac{\varphi_o}{\varphi_1} .$$

The normal components of velocity are bound with the angles φ and θ

$$\frac{v_{1n}}{v_{on}} = \frac{tg \left(\varphi - \theta \right)}{tg \, \varphi} = \frac{\varphi_o}{\varphi_1} .$$

For the ideal gas this relation can be written in the following
form :

$$\frac{tg(\varphi-\theta)}{tg\,\varphi} = \frac{(\gamma-1)\,M^2\sin^2\varphi+2}{(\gamma+1)\,M^2\sin^2\varphi}\,.$$

Here we use the law of conservation for the ideal gas when $M = M_o$
$\sin\varphi$. The equation can be written in the form :

$$tg\,\theta = 2\,ctg\,\varphi\,\frac{M^2\sin^2\varphi}{M^2(\gamma+\cos^2\varphi)+2}\,.$$

This equation is called the shock polar equation. If the gas is
not ideal, we must use the laws of conservation, as it was shown
above, and calculate the shock polar for the real gas. The ex-
citement of internal degrees of freedom of molecules changes the
results of the calculation. For instance, the value of the maximal angle of the wedge θ when the attached shock can exist,
changes in CO_2 ($V_o = 1900$ m/sec) from $42°$ in the equilibrium
case to $18°$ in the frozen case.

The reflection of shock waves from an inclined
surface can produce two wave configurations, depending on the
reflection conditions. The simplest configuration is regular
reflection.

The shock wave, moving with speed U_o , is incident on the solid surface at an angle W_1 . In the coordinate
system moving with the wave the gas flows into the latter with

a velocity U_o case W_1 and is deflected through an angle I .
Since subsequently to this the flow should be parallel to the
wall, a reflected wave is formed which will turn the flow in the
opposite direction through the same angle II .

Specifying the incident angle ω_1 and the shock
wave strength defines uniquely the state of the gas in the region between the incident and reflected waves. The strength and
position of the reflected wave is determined by the equation of
the shock polar (curve). But, for a given angle of turning of
the flow the shock polar defines two different shock waves,
those of the weak and those of the strong family. As is shown
experimentally, the reflected wave belongs to the weak family
of waves. This means that in the limit, when approaching incident waves with infinitesimal strength, the strength of the reflected wave also tends to zero, and the angle of reflection
approaches the angle of incident, which is actually observed in
acoustics. When $\omega_1 \to 0$, the reflected wave of the weak family
is transformed into a wave onto a solid body.

It follows from the properties of the shock polar
that regular reflection is by no means always possible. For a
given incident-wave strength there exists a limiting angle ω_k
regular reflection is not possible when $\omega_1 > \omega_k$. For an infini-
te incident-wave strength this angle approaches arc sin $1/\gamma$ i.e.,
for gases with $\gamma = 1.4$, $\omega_k = 40$. For waves with infinitesimal
strength $\omega_k \to 90°$, i.e., regular reflection is possible for

any angles of incidence.

On reflection of shock waves ω_1 , the angle of incidence, is not identical with angle of reflection ω_2 , which may be both greater and smaller than ω_1 . For a some angle of incidence $\omega_1^* = 1/2 \arccos(\gamma - 1/2)$, $\omega_1 = \omega_2$. When $\omega_1 > \omega_1^*$, $\omega_2 < \omega_1$; when $\omega_1 < \omega_1^*$, $\omega_2 > \omega_1$. Angle ω_1 is independent of the strength of the incident wave.

In cases when regular reflection is impossible, the incident A and reflected AR waves move away from the wall (Fig. 1) and a third shock wave is formed ; it is termed the Mach wave A M . The gas near the wall passes only through one wave, while far from the wall it passes through both the incident and reflected wave. Consequently, there should exist a contact discontinuity, emerging from the "triple" point, i.e., line A T .

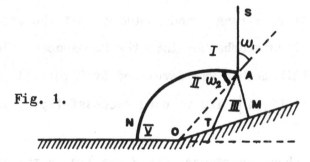

Fig. 1.

This wave reflection pattern is called Mach reflection (named after the physicist Mach who has first observed this phenomenon).

The Mach reflection phenomenon is so complex that up to present no theory is available for calculating, even in

the case of an ideal gas, the entire flow field and which would
have given satisfactory agreement with experimental results in
the entire range of Mach numbers of the incident wave and for
allangle of the reflecting surface.

Theoretical study of the reflection is difficult
not only because we are dealing with a nonlinear problem, but
also due to the fact that the entropy of the system is changing
and that the motion is of the vortex-flow type.

The first method of obtaining a solution consists
in linearizing the basic equations of motion on the assumption
that the incident shock is weak. In this case the change in entropy is a third-order infinitesimal as compared with the incident shock strength and it can be disregarded. This method was
first used by Sommerfeld [1] .

A second method consists in considering an incident shock of arbitrary strength and assuming that the obstacle
introduces only small disturbances into the homogeneous flow
behind the shock. This method was developed by Lighthill [2]
and Ting and Ludloff [3, 4] and is used successfully by many
other authors [5, 6].

The above approaches yield results in satisfactory agreement with experimentally obtained data [5, 7].

A more complex case of Mach reflection, i.e.,
that occuring when the variables of the medium undergo sharp
changes at relatively small distances (so-called short waves),

was studied by Ryzhov and Khristianovich [8, 9] . The case of
Mach reflection considered by them takes place at close-to-cri-
tical angles of incidence.

The majority of experimental studies of Mach
reflection pertains to weak waves with M < 3-4. Studies in this
range of Mach numbers have yielded the following relationships
for long shock waves [10-14] :

a) the triple point moves along a straight line
inclined at a constant angle χ to the reflecting surface ;

b) angle X for a constant-strength incident
wave increases with an increase in the angle of the reflecting
surface ;

c) the motion of the triple wave configuration is
self-similar.

The angles between waves in the Mach configura-
tion are determined satisfactorily from the three-shock theory.
In the general form this theory makes it possible to describe
the flow of the gas in all the angular regions which are formed
by the interacting waves. A detailed solution of the problem of
wave interaction can be found, for example, in [15] for the
steady-state case. For solving the problem of the Mach configuration, i.e., for the quasi-steady case, this theory was used
in [12, 16] . Let us consider the pattern of the Mach reflection
(Fig. 1). For this it is convenient to use a coordinate system
tied to the triple point, since in this system the reflection

can be regarded as quasi-steady. Then the triple configuration
can be examined without reference to the reflecting surface.
The location of shock waves in such a configuration will depend
appreciably on the direction of the incoming flow. In the coordinate system in which the triple point is at rest the direction of the incoming flow is determined by angle ω_1 .

The Mach reflection in the coordinate system tied
to the triple point is similar to regular reflection when the
latter is produced by a surface passing through the line of motion of the triple point perpendicular to the figure. By analogy
with regular reflection we shall also term angles ω_1 and ω_2
in the Mach reflection as angles of incidence and reflection,
respectively. In comparing with experimental results resort is
had to the fact that the angles of incidence is related to the
angle of motion of the triple point and the angle that the reflecting surface makes with the shock wave (wedge angle) by the
relationship

$$\omega_1 = 90° - (\chi + \alpha_0) \ .$$

Angles ω_1 and ω_2 in the Mach reflection are
related to angles of incidence and reflection on regular deflection by the expressions

$$\omega_1^M = \omega_1^R - \chi$$
$$\omega_2^M = \omega_2^R + \chi \ .$$

We now denote by ϱ, u, p, T, m and $\hbar$ the density, velocity, pressure, temperature, molecular weight and enthalpy of the gas, respectively , by 0 – the state ahead of the shock, by 1 the state behind the incident shock wave , by 2 the state behind the reflected wave , by 3 the state behind the Mach wave, and by 4 the state ahead of the Mach wave, which is the same as state 0 . The gas variables in region I are determined on the basis of velocity u_o by the method described earlier.

Let us consider a coordinate system tied to the triple point. In this system the gas "flows" into the incident wave with the velocity $u_o/\sin \omega_1$ at the angle $\varphi = \omega_1$. The incident wave deflects the flow passing through it by angle θ_1 . The value of this angle is determined from the continuity equation

$$\frac{\text{tg}\,(\varphi_o - \theta_1)}{\text{tg}\,\varphi_o} = \frac{\varrho_o}{\varrho_1} \; ; \qquad \theta_1 = \text{arc tg} \left[\frac{(1 - \varrho_o/\varrho_1)\,\text{tg}\,\varphi_o}{1 + \varrho_o/\varrho_1\,\text{tg}^2\,\varphi_o} \right] .$$

The velocity of the gas behind shock wave AS in the coordinate system tied to the triple point is

$$\bar{u}_o = \sqrt{u_1^2 + u_o^2\,\text{ctg}^2\,\varphi_o} \quad .$$

Let us consider the shock polar in coordinates p_2 , θ_2 for the reflected wave. Let the reflected wave be situated at angle φ_1 to the flow coming in with velocity $\bar{u}_1$, then the normal component of the flow velocity will be $\bar{u}_{1n} = \bar{u}_1 \sin \varphi_1$

We write the conservation equations at the reflec ted wave

$$\rho_1 \bar{u}_{1n} = \rho_2 \bar{u}_2 \qquad\qquad p_2 = \frac{\rho_2 R T_2}{m_2}$$

$$\rho_1 \bar{u}_{1n}^2 + p_1 = \rho_2 \bar{u}_2^2 + p_2$$

$$h_1 + \frac{\bar{u}_{1n}^2}{2} = \frac{u_2^2}{2} + h_2 \qquad\qquad \theta = \arctan\left[\frac{(1 - \rho_1/\rho_2)\tan\varphi_1}{1 + \rho_1/\rho_2 \tan^2\varphi_1}\right].$$

Solving this system by the same trial–and–error method as for the case of the incident wave, we will get for a number of values of $\bar{u}_{1n}$ values of p_2 and θ_2, i.e., the shock polar in the $p, \theta,$ plane, whose points represent possible values of the pressure ratio p_2/p_1 attendant to passing through the reflected wave and the deflection $\theta = \theta_2 - \theta_1$ of the flow in region 2.

Let us now construct the shock polar for the Mach wave. If this wave is situated at angle $\varphi_4 = 180^\circ - \omega_4$ to the incoming flow, then the normal component of the velocity to the Mach wave,

$$u_M = u_o \frac{\sin \varphi_1}{\sin \omega_1}.$$

The conservation equations at the Mach wave will be written as

$$\rho_o u_M = \rho_3 u_3, \qquad\qquad \rho_o u_M^2 + p_o = \rho_3 u_3^2 + p_3,$$

$$\frac{u_M^2}{2} + h_o = \frac{u_3^2}{2} + h_3 \; , \qquad\qquad p_3 = \frac{\rho_3 \, R \, T_3}{m_3}$$

Solving this system by the trial-and-error method, we find the values of p_3 and ρ_3 for the given u_M. The angle θ_3 of flow deflection is found from

$$tg \; \theta_3 = \frac{tg \; \varphi_4 \, (1 - \rho_o/\rho_3)}{1 + \rho_o/\rho_3 \, tg^2 \, \varphi_4} \; .$$

By specifying different values of u_M it is possible to construct the shock polar for the Mach wave in the coordinates $p_3 \, \theta_3$. In order to find the sought solution, one must satisfy the compatibility conditions which require that the pressures be equal and the flows be parallel in regions 2 and 3

$$p_3 = p_2 \; , \qquad\qquad \theta_3 = \theta_1 - \theta_2 \; .$$

These conditions are satisfied at the point of intersection of shock polars for the reflected and the Mach wave.

All the angles of the Mach configuration are determined from

$$\sin \varphi_1 = \frac{u_{1n}}{u_1} \qquad\qquad \omega_2 = \varphi_1 - \theta_1$$

$$\sin \varphi_4 = \frac{u_M}{u_o} \sin \omega_1 \; , \qquad \omega_4 = 180° - \varphi_4 \; .$$

It should be noted that the three-shock theory examined above
presupposes uniformity of flow in regions bound by the shock
waves. This means that : 1) all the intersecting discontinuities
are straight lines and 2) the flow parameters do not depend on
the distances from triple point A and that these variables are
constant in each of the regions bound by the discontinuities.

Thus, the use of the three-shock theory is possible if the angle of motion of the triple point is determined
experimentally, i.e., this theory is only a method for describ-
ing the flow pattern. Work done by Whithem [19] and Cabannes
[10] makes it possible to calculate, with some simplifying assumptions, the reflection of shock waves of any strength from
a wedge with an arbitrary wedge angle. Whithem has based his
deliberations on the following assumptions. The disturbances in
the flow which are produced by some or other phenomena, for
example, change in the tube's cross section, are regarded as
waves propagating through the shock. These disturbances, which
increase the Mach number of the shock, can become a discontinuity in the same manner as a plane compression wave becomes a
shock. This shock, which propagates through the original shock
is called a double shock. Actually this is a discontinuity in
the Mach number and the inclination of the shock being considered. On Mach reflection this double shock is the trace of the
triple point.

For small angles θ it is possible to compare

this theory with the linear theories of Lighthill [2] , Ting and
Ludloff [3] . Comparison shows that Whithem's method is suitable
for strong shocks with $M > 2$; its application to weak shocks
requires caution.

A different calculation of Mach reflection of
shock waves of arbitrary strength is given by Cabannes [20] . It
is based on the following assumptions.

1. The mutual positive on waves in the vicinity
of the triple point can be determined by equations of shock
polars, i.e., it is possible to apply to each of the shocks the
laws of conservation satisfying compatibility conditions.

2. The configuration which is realized from among
all those possible which satisfy the compatibility conditions is
that in which the Mach wave is perpendicular to the reflecting
surface.

3. A basic requirement is that requiring all the
shock waves comprising the Mach configuration to be rectilinear.
This theory is basically the three-shock theory, supplemented
by the requirement specified in item 2. Item 2 orients the three-
wave configuration relative to the reflecting surface. The en-
semble of these assumptions thus makes it possible, for a given
wave strength and angle θ of the wedge to determine angle χ and
all the other angles near the triple point.

Cabannes has solved this problem for a constant γ
of the gas on passing through the shock, assuming it to be 1.4

ahead and behind the shock.

The Cabannes and Withem theories are in the
greates disagreement for small Θ , for large Θ they yield
practically identical results. In addition, unlike Withem,
Cabannes has an obtained a Θ_{max} above which regular reflection should ensue.

When dealing with increasingly higher incident-
wave strengths the problem arises of the effect of variable 1
on the mode of Mach reflection. We know of, for example, work
by White [22] where he has observed changes in the wave configuration on irregular reflection in CO_2 , for which excitation
of molecular vibrations and dissociation take place at sufficiently low temperatures. In this case one cannot use theories
for γ = const.

In the study of molecular processes in gases and
of their effect on the flow variables we have used a methodical
approach which consisted in calculating the relationship between
the gas variables on various assumptions regarding the molecular
state. In the region of variables where the effect of physical
processes on this relationship is quite appreciable, the relationship between the flow variables were determined experimentally. Experimental data were compared with theoretical results
and this produced a picture of the state of the gas molecules
under the given conditions.

When proceeding in the above manner, we have

first checked the accuracy with which the relationship between
the gas variables behind a normal shock, derived on the basis
of the laws of conservation on the assumption of a one-dimension
al and steady-state flow, is applicable to a real flow of gas
behind a shock wave in a shock tube. For this purpose we have
measured variables behind a shock wave in gases the molecular
state of which is known and we have established the scale and
character of deviations from ideal for a flow in a shock tube.
Results of these experiments were considered when analyzing the
results of experiments in a reacting gas.

Studies were made of certain gasdynamic processes
under conditions when equilibrium state of the gas is reached
during a finite time period. For this we have studied characteristic and at the same relatively simple gasdynamic processes,
i.e., nonsteady propagation of a reflected shock wave in a gas
undergoing relaxation and the propagation of small disturbances
in a supersonic flow. Analysis of these processes has enabled us
to obtain data on the relaxation times for dissociation and vibrations of carbon dioxide and its mixtures with nitrogen at
temperatures of 2500-6000°K and on the structure of a small disturbance in a gas undergoing relaxation.

The data obtained on the effect of physicochemic-
al transformation on the normal reflection of a shock wave made
it possible to analyze a more complex gasdynamic phenomenon, i.e.
nonregular reflection of a shock wave on oblique incidence (the

so-called Mach reflection) in a real gas at high temperatures.
An experimental study of this phenomenon has shown that the real
properties of the gas result in a principally new system of
shock waves, i.e., a system with a double configuration attendant to Mach reflection.

REFERENCES

[1] Sommerfeld, A.Z. Math. Phys., Vol. 49, p. 11, 1901.

[2] Lighthill, M.I. Proc. Roy. Soc., Vol. 198, No. 1055, p.
         454, 1949.

[3] Ting, L. and H.F. Ludloff. JAS, Vol. 19, No. 5, p. 317
         1952.

[4] Ting, L. and H.F. Ludloff. JAS, Vol. 18, No. 2, p. 143
         1951.

[5] Fletcher, G.H., A.H. Taub and W. Blackney. Rev. Mod.Phys.,
         Vol. 23, No. 3, p. 271, 1951.

[6] Chester, W. Phil. Mag., Vol. 7, 45, 1293, 1954./Russian
         transl. In Mekhanika, No. 3, 17, 1954/.

[7] White, D.R. JAS, Vol. 18, No. 9, p. 633, 1951.

[8] Ryzhov, O.S. and S.A. Khristianovich. PMM, Vol. 22, Issue
         5, 1958.

[9] Grib, A.A., O.S. Ryzhov and S.A. Khristianovich. PMTF,
         No. 1, 63, 1960.

[10] Smith, G.L. Phys. Rev., Vol. 60, No. 11-12, p. 678, 1946.

[11] Taub, A.H. and G.L. Smith. Phys. Rev., Vol. 60, No. 11,
         p. 671, 1946.

[12] Bleakney, W. and A.H. Taub. Rev. Mod. Phys., Vol. 21,
      No. 4, p. 584. Russian transl. in VRT 1951, No.1.

[13] Bleakney, W.,C.M. Fletcher and D.K. Weimer. Phys. Rev.,
      Vol. 76, p. 323, 1949.

[14] Polachek, H. and R.J. Seeger. Proc. Symposia in Appl.
      Math., Vol. 1, p. 119, 1949.

[15] Courant, R. and K.O. Friedrichs. Supersonic Flow and
      Shock Waves. Wiley (Interscience), 1957.

[16] Polacheck, H. and R.J. Seeger. Phys. Rev., Vol. 84, No. 5,
      p. 922, 1951.

[17] Smith, L.G. Bull. Amer. Phys. Soc., Vol. 2, No. 4, p. 216,
      1957.

[18] Kawamura, R. and H. Saito. J. Phys. Soc. Japan, Vol. 11,
      1956.

[19] Whitem, G.W. J. Fluid. Mech., Vol. 2, pt. 2,No. 5, p. 548,
      1957.

[20] Cabannes, H. Lois de la Réflection des Ondes de choc dans
      les écoulements plans non-stationnaires. ONERA,
      No. 8, 1955.

[21] Bryson, A.E. and R.W.F. Gross. J. Fluid Mech., Vol. 10,
      P. 1, 1961.

[22] White, D.R. An Experimental survey of Mach reflection of
      shock waves, Proc. 2-nd Mid-West conference of
      Fluid Mech. p. 253, 1952.

[23] Bleakney, W. Proc. Symposia in Appl. Math., Vol. 5, p. 41,
      1954.

[24] Syshchikova, M.P., A.N. Semenov and M.K. Berezkina. In
      the collection Aerodinamicheskiye issledovaniya
      sverkhzvukovykh techeniy /see (96)/. Nauka
      Publishing House, Moscow-Leningrad, 1967.

[25] Gvozdeva, L.G. and O.A. Predvoditeleva. Dokl. AN SSSR,
         Vol. 163, No. 5, 1088, 1965.

[26] Gvozdeva, L.G. and O.A. Predvoditeleva. In the collection
         Issledovaniya po fizicheskoy gazodinamike /see
         (77) / , 183. Nauka Publishing House, Moscow,
         1966.

2. EXPERIMENTAL STUDIES OF THE BOUNDARY LAYER EFFECT ON DISTRIBUTION OF THE FLOW PARAMETERS BEHIND THE SHOCK WAVE IN A SHOCK TUBE

Introduction

According to the ideal theory of the shock tube, the flow parameters behind the shock front keep constant up to the contact surface, moving with a velocity equal to the gas velocity behind the shock wave. A closer examination of the prophile of the flow parameters behind the shock wave has revealed that these parameters vary from point to point along the plug. In particular, the parameters, determined from the conservation laws for an ideal gas, obtain only at the shock wave front, even though no physical and chemical transformations in the flow occur.

Indeed, all along the gas plug, there is a growth of the boundary layer on the tube walls. This phenomenon involves parameter variation in the nonviscous flow core as well.

The interaction of the nonviscous flow core with

the boundary layer in a shock tube has been calculated theoretically by several authors. A most complete analysis of the problem is to be found in the known studies of Mirels.

As it was shown by Mirels, the flow between the
shock wave and the contact surface in the shock tube can be described similarly to the flow in the nozzle with the area relation, A_{rs}/A_2 , which depends on the law of the boundary-layer
development. If l_m is the maximum value of the distance between
the shock and the contact surface, and the parameters on the
shock surface are denoted by index s , the parameters of the
flow [2] at the distance l from the shock are given by the relations :

$$M_2 = \left[1 - \left(\frac{l}{l_m} \right)^{1/2} \right] M_{2s} \left[\frac{2 + (\gamma - 1) M_2^2}{2 + (\gamma - 1) M_{2s}^2} \right]^{\frac{\gamma + 1}{2(\gamma - 1)}}$$

$$\frac{T_2}{T_{2s}} = \left(\frac{\rho_2}{\rho_{2s}} \right)^{\gamma - 1} = \left(\frac{p_2}{p_{2s}} \right)^{\frac{\gamma - 1}{\gamma}} = \frac{2 + (\gamma - 1) M_{2s}^2}{2 + (\gamma - 1) M_2^2}$$

$$t_f = \int_0^l \frac{\lambda l}{u_2} \qquad\qquad \bar{t} = \frac{l}{u_s}$$

$$\frac{\rho_s}{\rho_{2s}} \frac{t_f}{\bar{t}} = \left[\frac{2}{l/l_m} \right] \left\{ l_n \left[1 - \left(\frac{l}{l_m} \right)^{1/2} \right]^{-1} - \left(\frac{l}{l_m} \right)^{1/2} \right\} .$$

In particular, the problem of nonuniformity of
the flow in the shock wave plug in the shock tube has been considered by the same author in [1] . The plug nonuniformity estimate, given in [1] , is based on a few assumptions which
are valid only in the extreme cases. The first assumption is
that the shock front velocity (U_Λ) is to be constant :

(2.1) $U_\Lambda = const$

Further, U_s is assumed to be constant for such a long portion
of the tube (up to the point, where theoretical estimates are
taken), that the effects, related to the history of the initial
nonstationary wave propagation can be safely ignored. It is also
supposed, that the shock wave propagates with "maximal plug
length", provided the contact surface velocity ($U_{2,c}$) is equal
to U_Λ

(2.2) $U_{2,c} = U_{2,\bar{c}} \equiv U_\Lambda$

Even under these conditions, the distribution of
flow parameters along the plug, as obtained by Mirels, is appro-
ximate, and its accuracy is the better the higher the shock wave
velocity ; that is,it is supposed, moreover, that

(2.3) $\frac{1}{M_\Lambda} \ll 1$,

where M_Λ is Mach number of the shock wave. However, when the
shock wave velocity is very high, this solution becomes no long-

er valid, as it was arrived at under the assumption of constant

gas heat capacity and gas composition behind the shock wave,

whereas the assumptions of ideal stationary state (2.1-2.2) in

real flows in the shock tube are satisfied but approximately.

Thus, the assumptions underlying the study [1]

in a real flow in the shock tube are fullfilled only approxima-

tely. On the other hand, there are still few studies, devoted

to experimental research of nonuniformity of the plug, due to

the above interaction effect (of these, note [2-4]). Therefore,

we considered it desirable to carry out experimental determina-

tion of distribution of certain flow parameters behind the shock

wave in nitrogen, with Mach number (M_A) ranging from 3 to 7.

Experimental Technique

The experiments were performed in the shock tube

of the $4 \times 4 \, cm^2$ square cross section. The distance between the

test section and the diaphragm was 65 calibres, and the section

was within visibility of a Tepler device.

The parameters to be measured were flow velocity

(U_2) and flow Mach number (M_2). The selection of these very

parameters was due to the following reasons. On the one hand, it

is these parameters that are most sensitive to boundary layer

presence. On the other, they are but little affected by real

nonstationarity of the flow, and therefore, the comparison of

these very quantities with calculations according to the Mirels

theory is most meaningful. Here the authors made use of the
method of measuring the flow velocity by recording the propagation of an artificial density inhomogeneity ("Schliere"), produced by an electric discharge [5] in the flow behind the
shock. The discharge was produced by break-down of the gap between
the electrode on the tube wall and the grounded surface of the
semi-wedge, placed in the flow path within the shock tube 2 cm
from the tube wall.

The optimum energy of the discharge was chosen
($C = 50\,pF$, $U = 50\,kv$), so as to visualize the nonuniformity
density plug by Tepler method and at the same time to produce a
sufficiently weak density disturbance, so that the region in-
cluding the inhomogeneity should move with practically the
same velocity as the main flow. As is known, the region contain-
ing a density disturbance generally propagates with a speed
equal to that of the main flow only at the initial moment. Later
on, the nonuniformity may either lag behind or overtake the main
flow, depending on the sign and the amplitude of the density
disturbance. In the present experiments the observation time
was short enough (20μ sec) and, as shown by experiment, no
schlieren velocity change was observed during the time.

Therefore, the error in velocity measurement was
primarily due to the apparatus one, which included the inaccura
cy in measuring the inclination angle of schlieren trace in scanning
and the inaccuracy in determining drum revolution. The total

error amounted to ±2 per cent. Mach number was determined
from the inclination angle of Mach line developing on the semi-
wedge and particularly from the Mach line trace on scanning.

The accuracy of flow Mach number determination
from the inclination angle amounted to ±2 per cent, whereas
that from scanning to ±3 per cent.

Results

Measurements were made of the values U_2 and M_2
behind the shock wave front in N_2, with a shock Mach number (M_s)
ranging from 3 to 7 and the initial pressure (P_1) in the low
pressure chamber being about $20\,mm\,Hg$.

As a result of the experiments the values of U_2
and M_2 were found to exceed those calculated from conservation
laws ignoring the boundary layer effects. The calculated values
U_2, M_2 correspond to the assumption of fully-frozen vibrations [6]. The comparison between the "proper" time of a gas
species (at the moment of measurement) and the values of vibrational relaxation as measured in [7-8] shows, that under present
experiments, the assumption of complete vibrational freezing
holds true for most part of the plug.

Figs. 1-2 give some values of U_2 and M_2 measured
all along the shock wave plug. The theoretical curves in Fig. 1-
2 were calculated according to Mirels's scheme [1]. According
to the scheme, any physical flow parameter f is known to be a

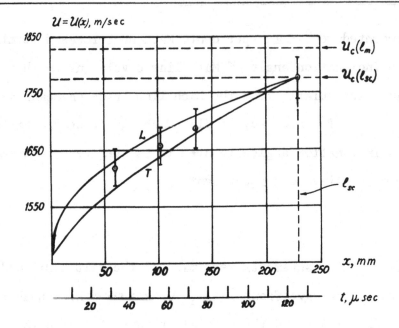

Fig. 1.

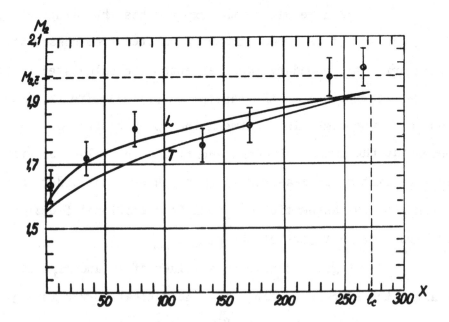

Fig. 2.

function of $X = x/l_m$, where l_m is "plug maximal length", and
x is coordinate along the plug, that is

$$\mathcal{L} = \mathcal{L}(X), \qquad X = x/l_m . \tag{2.4}$$

Similar to [5], we did not calculate the value of l_m and took
the plug length value directly from experiment, introducing a
correction for finiteness of value $\Delta_{\wedge c} U$:

$$\Delta_{\wedge c} U = (U_c - U_{2c}) \neq 0 . \tag{2.5}$$

That is, it is assumed here, just as it was in the original stud-
ies by Mirles (as[9]), that provided $\Delta_{\wedge c} U/U_{\wedge}$ is small, the profile
of any $\mathcal{L}$ (in the shock reference frame) is the same as for the
shock phase with "plug maximal length", the only difference being that, due to (2.5), $l_c > l_m$, where l_c is the contact surface
coordinate. If the theoretical value of $U(X_c)$ (where $X_c = l_c/l_m$)
is now equated to the contact surface velocity U_c measured from
experiment :

$$U(X_c) = U_c \tag{2.6}$$

then, given (U_c, X), it is possible to determine l_m, which
provides all necessary for calculation. It goes without saying,
that such a procedure of normalizing U distribution to a known value of U_c is not equivalent to theoretical calculation of l_m .
                    The present study did not specify whether the
boundary layer was laminar or turbulent. The information availa-

ble in literature indicates that a considerable section of the plug is expected to have turbulent boundary layer (e.g. 10). Consequently, two calculated curves are retained in figure 1, the ones corresponding to laminar ($\mathcal{L}$) and the other to turbulent (T) boundary layers, respectively; beginning just behind the shock wave.

It may seem paradoxical, that the calculated flow velocity increases with x growth more steeply for a laminar boundary layer, than for a turbulent one.

In fact, however, it should be remembered, that both the curves refer to the some plug length ; therefore, the curves are in a sense not comparable. For example, if the turbulent boundary layer develops, the laminar boundary layer with such short plug cannot exist at all.

Conclusion

The comparison of the experimental values of U_2 and M_2 along the shock wave plug in N_2, for $p_0 = 20$ mm Hg and $3 \leqslant M_s \leqslant 6$ with data calculated by Mirels'theory [1] , shows a fair agreement between the two. That is, for most of the plug, except probably for the immediate vicinity of the contact surface, the experimental values (Figs. 1, 2) agree with calculation within an accuracy of ± 3 % (i.e., the greatest departure from calculated curves, corresponding to laminar or turbulent boundary layers does not exceed 3%).

More accurate comparison with the theory could be accomplished, if the point of transition of laminar boundary layer to turbulent one were known.

The present study, however, was not designed to determine the point of transition by experiment. Moreover, the plug length was also taken from experiment, not from calculation. Nevertheless, comparing the approximations underlying the calculation with the experiment accuracy of flow parameters (U_2, M_2) the agreement between theory and experiment achieved may be considered satisfactory.

REFERENCES

[1] H. Mirels, Flow nonuniformity in shock tubes operating at maximum test times, Phys. Fluids, $\underline{9}$, 1907, (1966).

[2] J.N. Fox, T.L. McLaren, R.M. Hobson, Test time and particle paths in low-pressure shock tubes, Phys. Fluids, $\underline{9}$, 2345 (1966).

[3] M.P. Syshchikova, A.N. Semenov, M.K. Berezkina. Measurement of Flow-Parameters in Shock Tube. Collection of papers. "Aerophysical Studies of Supersonic Flows", Nauka, Moskva-Leningrad, 1967.

[4] W.H. Christiansen, Measurement of flow uniformity in shock tubes operating at low initial pressure, Phys. Fluids, $\underline{11}$, 2279 (1968).

[5] R.G. Nemkov. Flow velocity as measured in Shock Tube by Spark Technique". Collection of Papers "Physical Gas Dynamics of Ionized and Chemically reacting Gases", Nauka, Moskva, 1968.

[6]   I.M. Naboko, T.V. Bazhenova, R.G. Nemkov, Experimental
        studies of the parameters of the flow behind a
        shock wave moving in relaxing gas, 1-st interna-
        tional colloquium on gas dynamics of explosions,
        Astronautica Acta, 14,5, 497 (1969).

[7]   R.C. Millikan and D.R. White, Vibrational energy exchange
        between N_2 and CO - the vibrational relaxation
        of nitrogen, J. Chem. Phys., 39, 98 (1963).

[8]   I.R. Hurle, Line reversal studies of the sodium excita-
        tion process behind shock waves in N_2 , J. Chem.
        Phys., 41, 3911 (1964).

[9]   H. Mirels, Test time in low-pressure shock-tubes, Phys.
        Fluids, 6, 1201 (1963).

[10] J.C. Breeze and C.C. Ferriso, Duration of laminar flow in
        a shock wave boundary layer, Phys. Fluids, 7,
        1071, (1964).

3. FLOW PARAMETERS BEHIND THE SHOCK WAVE IN RELAXING GASES

At the high temperatures which are produced
behind shock waves, the gas molecules undergo physicochemical
transformations. The internal degrees of freedom of the molecules are excited, dissociation ensures, new compounds form and
the gas is ionized. All these processes affect the gasdynamic
variables of the flow behind the shock wave. The extent of the
effect of the physicochemical transformations on the flow variables of the gas is determined by the degree of molecular transformation. In the flow behind strong shock waves the degree of
molecular excitation is high and the equilibrium excited state

is reached rapidly, since the relaxation time of excitation processes at high temperatures is small. The gas variables reach their equilibrium values in fractions of microseconds.

However, each gas has a temperature range (and correspondingly, a range of values of M_o , the Mach number of the wave), for which molecular excitation has an appreciable effect on the flow variables, and at the same time the excitation process duration is of sufficient length, comprising tens and hundreds of microseconds. Under conditions which have corresponding to them large relaxation times the gas flow is determined by nonequilibrium variables. The flow structure under these conditions is highly variegated. This consists, in particular in the fact that at small distances from the source the small disturbances propagate at a speed equal to the high-frequency speed of sound [13, 14, 18, 115, 127] . The thermodynamic variables of the gas under these conditions can be determined by measuring the gasdynamic flow variables.

One of the variables which depends appreciably on the thermodynamic state of the flow is M_1 , the Mach number of the flow. The angle of inclination of a small disturbance in the flow, which is determined by the ratio of the flow velocity to the speed of sound, has different values depending on the extent of molecular excitation.

The time of relaxation of CO_2 molecules as determined by different authors is different [1-8] . Therefore, it is useful to obtain some more experimental data.

Results

References [9-12] contain the results of flow
parameter measurements behind shock waves in the shock tube.
The measurements were made in N_2, O_2 and CO_2 The measured parameters were flow Mach numbers, flow velocity and the velocity
of weak propagation, all of which were determined as a function
of Mach number of an incident shock wave M_0

The conclusions about thermodynamic state of the
gas behind the shock wave were made by means of the comparison
of experimental results with a number of possible calculated
values. Calculations were made for the flow state with certain
extent of molecular excitation.

Fig. 1 presents the following calculation curves
of M_0 dependence of M_1 in CO_2 :

1 - all the degrees of freedom are in equilibrium ($P_0 = 10\,mm\,Hg$)

2 - the dissociation is frozen, the vibrations
are in equilibrium state

3 - only two degrees of vibrations are excited
(ω_1 and ω_2), sound speed is in equilibrium

4 - the vibrations are in equilibrium, the
sound speed is frozen ($\gamma = 1.4$)

5 - only two degrees of the vibrations (ω_1 and

ω_2) are excited, the sound speed is frozen ($\gamma = 1.4$)

      6 - only one degree of the vibration (ω_1) is excited, the sound speed is frozen

      7 - vibrations of molecules are frozen, the sound speed is frozen.

      The values of M_1 in N_2 and in mixtures of N_2 and CO_2 were calculated in a similar way.

      It has been found that flow parameters in O_2 are in agreement with well established data of other authors, and correspond to the equilibrium state of the gas behind the shock. In CO_2 and N_2, equilibrium was not reached, no dissociation was found to occur and internal degrees of molecule freedom had no time to get fully excited. Analysis of the experimental data and estimation of their maximum possible deviation from calculation ignoring boundary-layer effect, reveals a difference in this deviation dependent on a particular gas. Mach numbers for nitrogen flow as measured by experiment prove to be somewhat exceeding the calculated values. In CO_2, experimental points were found to be somewhat below the curves calculated in neglect of boundary-layer effect (Fig. 2). The indexing of calculated curves in Fig. 2 is the same as in Fig. 1. Fig. 2 presents only the curves which are the best fit to experimental data.

      The plots of Fig. 3 and 4 present experimental points and calculation curves for flow Mach number in mixtures of CO_2 and N_2, with the $CO_2 - N_2$ ratio of 75 to 25 in Fig. 2,

and of 30 to 70 in Fig. 4, respectively. The content of water
vapours was up to 2 %.

Curves 5 in the figures correspond to calculation
made under the assumption of no asymmetric valency vibration in
CO_2 molecules, and no vibration in N_2 molecules.

When measuring flow Mach number in mixtures the
same tendency is revealed as found in the case of pure CO_2 and
N_2 , increase in N_2 content leads to enhanced values of M_2 as
compared to calculation. In mixtures with CO_2 being the prevailing component, experimental values are in better agreement with
calculation.

Interpretation of Results

If a correction term to allow for boundary-layer
effect on the flow parameters in the shock tube is introduced
into the calculation, the calculated values are expected to be
somewhat higher than 5 curves. At $M = 6$, maximum increase in
flow Mach number as calculated according to Mirels theory (13),
may amount to 5 - 6 per cent. With increase in M_o , relative
increase in flow Mach number is expected to be smaller. In (14)
an estimation is made of the boundary-layer effect on flow parameters in the shock tube, following the scheme proposed by
Mirels.

If the calculated Mach numbers in Fig. 2-4 are
correspond for the boundary-layer effect as estimated in [14],

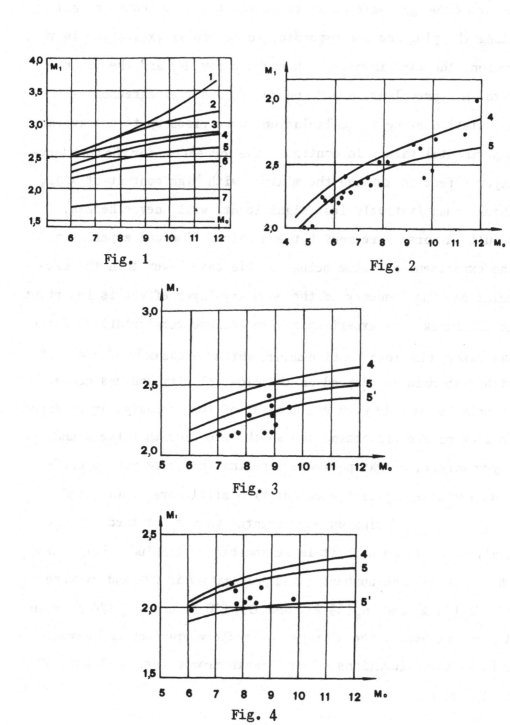

Fig. 1

Fig. 2

Fig. 3

Fig. 4

assuming the gas heat capacity behind the shock constant all
along the plug and corresponding to molecular excitation in this
region, the experimental Mach numbers for N_2 and the mixture
with N_2 prevalent, are found to be in better agreement with
calculation since the calculation curve rises up to an "average"
experimental value. In contrast, the correction for the boundary-
layer effect in CO_2 and the mixture with high content of CO_2,
though quantitatively less significant, will, nevertheless,
result in worse agreement between calculation and experiment,
the experiment at value being in this case lower than the calculation. The tendency of the boundary-layer effect is important
in all shock tube experiments also because recalculation of the
laboratory time into the time during which the gas molecules are at
high temperatures, depend on the flow velocity and has consequently to be made with due account of the boundary-layer effect
on flow parameters behind the shock. The fact that the boundary-
layer effect, on gas parameters behind the shock may be different for gases makes the calculation still more complicated.

Basing on experimental values obtained and the
analysis of these data it is reasonable to conclude that behind
the shock at Mach numbers $5 \div 11$ propagating in CO_2 and mixtures
of CO_2 (75 %) and N_2 (25 %), and CO_2 (30 %) and N_2 (70 %) with
2 per cent Steam, the nitrogen molecule vibrations and asymmetric valence vibrations of CO_2 remain frozen for, at least, 100
to 150 μ sec.

The accuracy of the information concerning the gas all along the plug in gas-dynamic experiments, as well as in any other experiments in shock tubes, is limited by our ability to take account of the boundary layer effect on the flow parameters behind the shock.

The tendency of this effect as shown by experiment is different with different gases and cannot be taken into account by means of the theories of this phenomen, available up to date.

REFERENCES

[1] Griffith W.C. In the book "Fundamental Data obtained from Shock Tube Experiments", ed. Ferri A., p. 242, Pergamon Press (1963).

[2] Zinkiewicz H., Iohannesen N., Gerrad J. Further Results on the over-all density ratios of shock waves. J. Fluid Mech., 17, 267-270 (1963).

[3] Michel K.W., Z.f. Phys. Chem., v. 44, 160 (1965).

[4] Gaydon A.G., Hurle I.R. The Shock Tube in High-Temperature Chemical Physics. Chapman and Hall, London (1963).

[5] Hurle J.R. Time-Reversal Studies of the Sodium Excitation Process behind Shock Waves in N_2, Journ. of Chem. Phys., 41, No. 12, p. 3911 (1964).

[6] Millican R.C., White D.R. Systematics of Vibrational Relaxation. Journ. Chem. Phys., 39, 3209 (1963).

[7] Sebeacher D. A Correlation of N_2 Vibrational–Translational Relaxation Times. AIAA J. 5, No. 4 (1967).

[8] Taylor R.L., Bitterman S. Vibrational energy transfer in
     the $CO_2 - N_2 - H_2O$ molecular system. VII International Shock-tube Symposium, Toronto, June (1969).

[9] Naboko I.M., Bazhenova T.V., Namkov R.G., Experimental
     Studies of the Parameters of the Flow behind a
     Shock wave Moving in Relaxing Gas. Astronautica
     Acta, 14, 497–502 (1969).

[10] Bazhenova T.V., Grozdeva L.G., Lobastov Yu.S., Naboko I.M.
     Nemkov R.G., Predvoditeleva O.A. Shock Waves in
     Real Gases, Moscow, Nauka Press, (1968), (translated : NASA TTF–585).

[11] Naboko I.M. Investigation of the flow state behind the
     shock wave by observing the flow pattern develops when overflowing the obstacle in the shock
     tube. In "Investigations on Physical Gas Dynamics"
     Nauka, Moscow (1966).

[12] Bazhenova T.V., Naboko I.M. About the rate of the physical-chemical transformations of CO_2 molecules
     behind the shock wave by the temperatures of 2000–
     4000°K, Dokl. Acad. Nauk USSR, 154, 401–403
     (1964).

4. UNSTEADY PROPAGATION OF A REFLECTED SHOCK WAVE
IN A REACTING GAS

If the reactions taking place behind a reflected
shock wave are nonequilibrium, then their development with time
may introduce quantitative features into the process of reflection of the wave. A reaction which does not succeed in reaching
equilibrium in gas regions near the front of the reflected
shock wave develops in gas regions near the end. These are heated by the reflected wave before others. This produces changes
in time in the boundary condition at the reflecting surface, for
which reason on the propagation of the reflected wave becomes
unsteady. The study of propagation of reflected waves in reacting gases is of great interest from this point of view. In addition, measurement of flow variables behind reflected shock
waves and the velocity of propagation of the reflected wave may
yield information on the total rates of relaxation processes
which take place in the test gas at elevated temperatures.

The relationship between the relaxation time
and the time of reaching a shock-wave velocity close to that on
equilibrium was obtained by Spence [1] . He has considered the
unsteady-state propagation of a shock wave ahead of a piston in
a gas undergoing relaxation by the method of linearized charac-
teristics. The problem of reflection of a shock wave from a flat

wall differs from that of propagation of a wave ahead of a piston by the fact that the velocity of the piston is assumed to be
zero, the initial flow comes in with the velocity of the piston.
In solving the problem of propagation of a shock wave ahead of a
piston, Spence examined the equations of motion of a gas undergoing relaxation in the lagrangian

coordinates $\quad h = \int_{x(o,t)}^{x(h,t)} \rho \, dx :$

$$\frac{\partial u}{\partial t} = - \frac{\partial P}{\partial h}$$

(4.1)
$$\frac{\partial u}{\partial h} = - \frac{\partial}{\partial t} \left(\frac{1}{\rho} \right)$$

$$d \frac{dE}{dt} + P \frac{\partial}{\partial t} \left(\frac{1}{\rho} \right) = 0 .$$

The equation in the characteristic form is written as

$$\frac{a}{\gamma P} \left(\frac{\partial}{\partial t} \pm \rho a \frac{\partial}{\partial h} \right) P \pm \left(\frac{\partial}{\partial t} \pm \rho a \frac{\partial}{\partial h} \right) u = - \frac{E}{a} \frac{d\alpha}{dt}$$

where a is the frozen speed of sound and α is the degree of
dissociation of the gas

$$a^2 = \gamma(\alpha) \frac{P}{\rho} .$$

The reaction rate $d\alpha/dt$ can be obtained if the
relationship $\alpha(t)$ is known. If the reaction does not introduce
appreciable changes into the state of the medium, then

$$\alpha(t) = \alpha_0 \, th \, \frac{t-t_0}{2\tau_0} ,$$

where t_0 is a constant of integration, τ_0 is a constant characterizing the time for reaching degree of dissociation α_0 which is close to that at equilibrium.

It is assumed in the solution that the boundary conditions which are satisfied by the variable flows immediately behind the shock waves are applicable at the undisturbed path of the wave $X = U_0 t$ where U_0 is the shock wave velocity, if there is no dissociation between the wave and the piston. Then the Lagrangian coordinate $h_\Lambda(t)$ of the shock wave, equal to the mass of the gas which flowed through the shock wave during time t, is

$$h_\Lambda(t) = \varrho\left(U_0 - u_0\right)t ,$$

where u_0 is the piston velocity.

After substituting the expression for α and linearization, Eq. (4.2) takes on the form

$$\left(\frac{\partial}{\partial t} \pm \varrho_0 a_0 \frac{\partial}{\partial h}\right)\left(\frac{P}{\gamma P_0} \pm \frac{u}{a_0}\right) = -\left(\frac{\alpha_0}{2\tau_0}\right)\left(\frac{E_0}{a_0^2}\right)^\ell c\, h^2 \left\{\frac{t-t_0(h)}{2\tau_0}\right\}. \quad (4.2)$$

The general solution of this equation

$$\frac{U-u_0}{a_0} = A\left[f\left(t - \frac{h}{\varrho_0 a_0}\right) + q\left(t + \frac{h}{\varrho_0 a_0}\right) - t\, h\left\{\frac{t-t_0(h)}{2\tau_0}\right\}\right], \quad (4.3)$$

where f and g are arbitrary functions, while $M = \dfrac{U_o - u_o}{a_o}$,

$$(4.4) \quad A = -\frac{a_o E_o}{a_o^2}\,\frac{M}{1-M^2}\,\frac{P-P_o}{\gamma_o P_o} = A\left[f\left(t-\frac{h}{\rho_o a_o}\right)-g\left(t+\frac{h}{\rho_o a_o}\right)-M\,\mathrm{tg}\left\{\frac{t-t_o(h)}{2\,T_o}\right\}\right].$$

When boundary conditions $h = h_\lambda(t)$ are substituted into Eqs. (4.3) and (4.4), the values of u and p become equal to these gas variables immediately behind the shock, when these are determined on the basis of the conservation laws. The use of this boundary condition across the shock makes it possible to eliminate the arbitrary function   , while function is found from the boundary condition at the piston where $u = u_o$ for all the t .

The expression for $f(t)$ is given by the functional equation

$$f(t) + \nu f(\lambda t) = \mathrm{th}\,\frac{t}{2\,T_o}$$

where ν and λ are functions of the Mach number of the flow.

For the case of a shock wave reflected from a rigid wall ($u_o = 0$), the Mach number of the flow behind the wave is u_o/a_o . When passing through the front of the reflected wave   may be regarded as unchanged, since the active degrees of freedom have already been excited in the incoming flow behind the incident shock wave, while the unactive degrees of freedom are not excited immediately behind the shock due to lack of time. Analyzing the boundary conditions for this case we get the following values of λ and ν

$$\lambda = \frac{1-M}{1+M}$$

$$(4.5)$$

$$\nu = \left[\left\{ 1 - 2\left(\frac{\gamma-1}{3-\gamma}\right)M \right\} \Big/ \left\{ 1 + 2\left(\frac{\gamma-1}{3-\gamma}\right)M \right\} \right] \lambda^2 .$$

The time needed for the velocity to reach its limiting value is, by definition, infinity ($t=\infty$, $U=U_\infty$), however, analysis of functional equation (4.5) performed in [18] shows that $f(t)$ come quite close to the equilibrium value during a time $\tau_\Delta = \int_0^\infty |1 - F(t)| \, dt$. This time is a function of τ_0, the relaxation time of the reaction

$$\tau_\Delta = \tau_0 \frac{2 \ln 2}{1-M^2} \left\{ \frac{1 + \frac{3\gamma-1}{3-\gamma} M^2}{1 + 2\frac{\gamma-1}{3-\gamma} M^2} \right\} .$$

$$(4.6)$$

Analysis of experimental and theoretical data on the reflection of shock waves shows that, if the time of the relaxation process is commensurable with the time of propagation of the wave, then it is possible to observe nonsteady propagation of the reflected wave and to determine the relaxation time by the relationship governing the deceleration of the reflected wave. Here it is necessary to take into account corrections introduced into the reflected wave's velocity by the energy losses occurring on reflection.

Comparing the velocities of a reflected wave in CO_2 for initial pressures p_0 of 3 and 12.7 mm of Hg, on different assumptions on the thermodynamic state of the gas behind the incident and reflected shock waves, we see that the velocities obtained on the assumption of complete equilibrium (1) differ appreciably from the values calculated without taking molecular dissociation into account (2), starting with M = 8 of the incident wave, to values calculated on the assumption of a constant ratio of specific heats ($\gamma = 1,3$) when passing through the shock (3), starting with M = 4

To discover the effect of nonequilibrium processes in the gas on the velocity of a reflected shock wave, experiments were performed in CO_2 for M = 4-12 of the incident wave with an initial pressure $p_0 = 12.7$ Hg .

The results of measuring the velocity of the reflected shock wave are presented in Fig. 1. Curve 1 corresponds to the case of equilibrium dissociation of the gas behind the incident and reflected shock waves ; curve 2 of $c_p(T)$ is calculated on the assumption of equilibrium excitation of molecular vibrations, without considering dissociation ; curve 3 ($\gamma = 1.3$) assumes that the gas is ideal. The experimentally determined velocities of the reflected wave for M < 8 lie near the curve of $c_p(T)$,while for M > 8 the velocities measured directly at the tube's end lie near this curve and below it, while the velocities after deceleration lie near the curve

corresponding to complete equilibrium in the system, and below.
The fact that the velocity of reflected waves after deceleration obtained in some experiments is below the equilibrium velocity is attributable to energy losses on reflection. The values of D_2 on assumption of losses are lower than the ideal
theoretical values by 10 %. Hence it cannot be claimed that the
velocity after deceleration corresponds to the state of total
equilibrium of the gas behind the reflected wave. The measured
time between the instant of reflection and the start of deceleration points only to the fact that the dissociation reaction
at the tube end has not as yet reached the equilibrium extent,
i.e., the relaxation time of the gas behind a reflected wave is
greater than that measured.

       We now estimate the shortest time during which
a change in the state of the gear near the butt can effect the velocity of the reflected wave. Then the time needed for establish
ing the equilibrium velocity of the reflected wave can be used
to determine the time of establishing the equilibrium concentrat
ed in the gas layer near the butt, which was the first to be
stopped by the reflected wave.

       Let the composition of the gas at the butt reach
its equilibrium value during a time τ_0 after reflection of the wave
which propagates relative to the butt with a velocity D_2 .
When the first signal about the changed state at the butt, which
moves with velocity a, will reach the reflected front at the

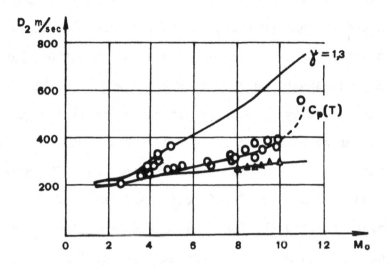

Fig. 1

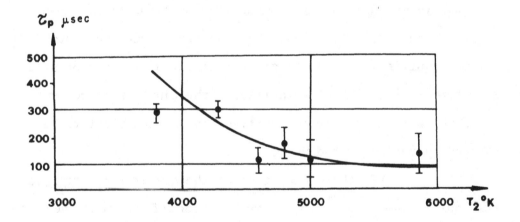

Fig. 2

time τ_{Λ} , which is equal to $\tau_0 + t_1$ (t_1 being the time need-ed by the signal to overtake the reflected shock). The relation-ship between these times can be easily obtained by equating the expressions for the front's coordinates at the time of meeting τ_s

$$D_2 \tau_{\Lambda} = a_2 t_1$$

$$\tau_{\Lambda} = \tau_0 + t_1 .$$

Then the time needed for establishing the equilibrium concentra-tion will be expressed in terms of the time needed by the first signal to reach the reflected shock as follows

$$\tau_0 = \tau_{\Lambda} (1 - D_2/a_2); \quad D_2/a_2 = M .$$

The maximum signal velocity a_2 is apparently equal to the frozen speed of sound under conditions prevailing behind a re-flected shock wave. The experimentally determined values of τ_s range from 30 to 7.5 μsec (it should be noted that due to the fact that this transition is smooth, the time is measured to approximately within 20 %). These data were used for calculating τ_0 , as well as $\tau_0 p$ reduced to atmospheric pressure.

The relationship between the relaxation time and the time for obtaining equilibrium velocity of a reflected wave can be determined more exactly from Eq. (4.6) derived by Spence. This formula uses the ratio of specific heats γ , which has different values across the front and near the piston. However, calculations using this expression show that the ratio τ_{Λ}/τ_0

depends little on the value of γ when the latter is in the range from 1.3 to 1.15.

Applying Eq. (4.6) to the measured times for establishing equilibrium velocity of the reflected wave in CO_2 we see, that τ_0 the dissociation relaxation time is by approximately 60 % lower than the time τ_A of establishing equilibrium velocity of the wave.

The values of τ_A/τ_0 determined on solving the problem of flow of gas undergoing relaxation ahead of the piston, is only by 15 % higher than the value obtained from simple considerations concerning the propagation of elementary disturbances with the speed of sound from the butt to the shock wave ($\tau_A/\tau_0 = = 1/(1-M)$). The dependence of time of establishing equilibrium dissociation on the temperature of the gas behind a reflected shock is shown in Fig. 2. Here T_2' denotes the temperature across the reflected shock, calculated on the assumption of frozen dissociation and excited rotational and vibrational degrees of molecular freedom ; $\tau_0 p$ are the times of dissociation relaxation reduced to atmospheric pressure, recalculated from experimental values of τ_s using Eq. (4.7).

These results pertain to dissociation of CO_2 with a 2 % content of water vapor. Of interest for gasdynamic studies is the relaxation time in moist CO_2, since flows of this gas practically always take place in the presence of water vapor.

The relaxation of dissociation and vibrations at high temperatures has an appreciable effect on the flow of CO_2 and, in particular, on the state of the gas behind a shock which is produced in flows about bodies. The determined dissociation relaxation times can be used to determine the distances at which equilibrium variables of CO_2 are established behind a normal shock. The results obtained for a normal shock give an idea about the state of the gas behind a detached shock which is produced in the flow about a blunt body, for example, when a space ship enters the atmosphere of the planet Venus. At atmospheric pressure behind the shock, when the velocity of the incoming flow is 4kg/sec equilibrium behind the shock wave is established at a distance of 3 cm from the shock front, for a flow velocity of 3 kg/sec this distance is 10 cm, while at 2.5 km/sec equilibrium is not established at a distance greater than 15 cm.

The gas temperature in the above region ranges from T_2' at the front corresponding to frozen dissociation, equilibrium values. The difference between these temperatures is several thousand degrees, Fig. 3.

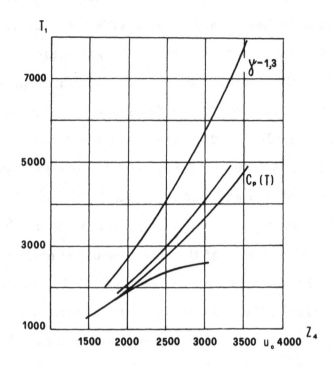

Fig. 3

REFERENCES

[1] D.A. Spens. Mekhanika, No. 6, 65, 1962.

5. MACH REFLECTION OF STRONG SHOCK WAVES IN REAL GASES

To determine the effect of real properties of a gas attendant to irregular reflection of waves, studies were made of the reflection of strong shock waves in CO_2, air, nitrogen and argon. In CO_2 excitation of molecular vibrations and dissociation are quite perceptible at sufficient low temperatures, when M_0 of the incident wave do not exceed 7-8. In air and nitrogen these effeffects at the same Mach numbers are less pronounced.

The experiments were performed with the UT-2 shock tube. A wedge with a vertex angle of α_0 was placed into the test section. In the majority of cases the wedge was raised above the lower wall of the tube. The process was visualized by the schlieren method. It was photographed by the SFR-L camera and the FR-185 photorecorder, which made possible obtaining time scanning of the process.

The vertex angles of the wedge were 10, 24, 28, 32 and 36°. The initial pressure p_0 ahead of the incident wave was usually 12.7 mm of $H_g = 1.68 \cdot 10^4$ dynes/cm^2.

The angles between all the waves comprising the Mach configuration were measured on the photographs and their variation in time was checked. The shock wave velocity was measured from the photographs and at the same time by means of piezoelectric pickups placed along the tube.

The photographs taken in CO_2 show clearly that the reflected wave has a break. A new shock wave is formed at the break, i.e., a second triple point is formed. The reflected wave is situated below the path of motion of the triple point. The Mach wave changes its inclination on passing from the triple point to the surface of the wedge (Fig. 1), [1-6] .

Mach reflection was observed during a short time interval, limited by the time during which the shock wave passed along the wedge. For a wave with $u_o = 2$ km/sec this time is not more than 30 μsec . During this time the process is self-similar, the accuracy being 3 %.

As was pointed out in Section 1, the available methods for calculating irregular reflection require the satisfaction of the following conditions : rectilinearity of waves emanating from the triple point, homogeneity of flow in the regions between them, perpendicularity to the reflecting surface of the Mach waves, as well as the requirement that the ratio γ of specific heats remain constant.

It is seen from photos that on reflection of a shock wave moving in CO_2 at $u_o = 2000$ m/sec the Mach wave is not a straight line. And even if its base is perpendicular to the surface of the wedge, these conditions are not satisfied near the triple point. Hence in our subsequent calculations of the Mach configuration we shall use experimental values of χ .

When applying the three-shock theory to the study

of wave configuration reference should be had to the formation
of an angular vortex region instead of the tangential discontinuity. It is shown that formation of this angular region shifts
the values of angles between the waves forming the triple configuration by 1–2°, which is within the limits of experimental error. Hence when using the three-shock theory we have considered
the tangential surface as the classical tangential discontinuity.

The physical and chemical transformations behind
the fronts of strong shock waves were taken into account as follows. The state of the gas and the location of shock waves in
the vicinity of point A were calculated on the basis of the
three-shock theory (Section 1), while the enthalpies of the gas
$H = H(T, p)$ were taken according to the assumed degree of
physical and chemical transformations. Since the time of exis-
tence of the entire configuration in our experiments was small
and commensurable with the relaxation times, calculations were
made on several assumptions.

Version 1. The gas behind the incident as well as
behind the reflected and Mach waves is in complete thermodynamic
equilibrium, i.e., molecular vibrations of the gas are excited
and dissociation takes place. The values of H and u under this
version were taken from tables.

Version 2. In the CO_2 and nitrogen molecular vibrations were excited while there was no dissociation. In air
the oxygen molecules not dissociated and molecular vibrations were

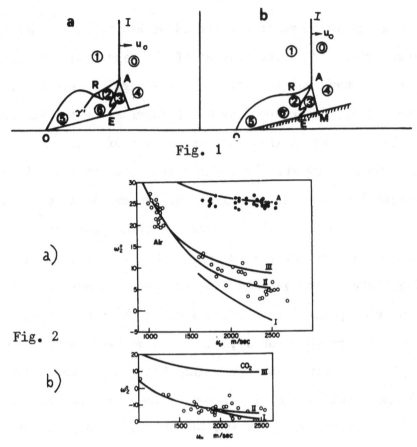

Fig. 1

a)

Fig. 2

b)

not excited in the N_2 . There was no dissociation in

Version 3. No physical or chemical transformations took place in the test gas. The ratio of specific heat
in the gas behind the wave remained the same as ahead of the
wave.

A comparison of experimentally determined reflec
tion angles ω_2 with those obtained from the three-shock theory
as a function of u_o , the incident-wave velocity for a constant
wedge angle $\alpha_o = 24°$ in argon, air and CO_2 is presented in
Fig. 2. The values of ω_2 as a function of u_o and of angle ω_1

were calculated theoretically. Angles ω_2 and $\omega_1 = 90 - (x + \alpha_0)$ for u_0 were obtained experimentally.

In argon (Fig. 2,a) the experimentally obtained values of ω_2 are in agreement with curve 3, since no physical or chemical transformations take place in argon under the conditions under study. In air in the range of incident-wave velocities of 900–1200 m/sec the points lie on curve 3. In the velocity range of 1800–2500 m/sec the relaxation time for exciting molecular vibrations in O_2 and N_2 and the dissociation time for O_2 is shorter than the time during which the gas is in the heated state, while the dissociation relaxation time for N_2 is much greater. The points cluster around curve 2. In CO_2 (Fig. 2,b) up to velocities $u_0 = 1800$ m/sec the experimental values of ω_2 lie near the curve taking into account excitation of internal degrees of freedom. Consideration of dissociation in high-velocity incident waves modifies the value of ω_2 within limits of experimental error. Due to the fact that one of the relaxation times is shorter and the other is longer than the process under study proper, the process of reflection in the velocity range under study is self-similar. It is probable that in the case of even higher velocities this self-similarity ceases to exist if the time during which the gas is in the hot state is comparable with the dissociation relaxation time.

In order to show the effect of physical and chemical transformations, it is convenient to use the concept of

effective ratio of specific heats γ_{ef}. The enthalpy of a gas
undergoing physical and chemical transformations is a complex
function of the temperature. Similarly to the case of a gas
with constant specific heat we can introduce the expression

$$h = \frac{\gamma_{ef}}{\gamma_{ef}-1} \frac{p}{\rho}.$$

where h is the specific enthalpy, p is the pressure and ρ is
the density. This approximation makes it possible to use for
complex calculations the Hugoniot /Shock/ relations, in which
we now use the effective ratio of specific heats. Calculations
were made for the triple configuration which is produced by a
wave incident on a wedge when the former moves with a velocity
of 1.9 km/sec for different values of γ_{af}.

Thus, reflection angle ω_2 is very sensitive to
variations in γ_{ef}. A moderate change in γ_{ef} appreciably increases the angle between the incident and reflected waves.
For $\gamma_{ef} = 1.15$, angle ω_2 is negative for all values of the angle of incidence. If the gas is assumed to be ideal ($\gamma = 1.66$
and 1.4) angles ω_2 are always positive.
Angles ω_2 are displaced toward the positive range.

Comparison of experimentally and theoretically
obtained values of reflection angles $\omega_2 = \omega_2(\omega_1, u_0)$ vali-
dates the application of the three-shock theory modified so as
to take into account variations in the γ of the gas for calcu-

lating irregular reflection of strong shock waves in the vicinity of the triple point. Since the pattern of wave diffraction
on the wedge is composed of Mach reflection and of supersonic
flow about the wedge angle, it is required that the reflected
wave become an attached wave.

Calculations made for a wave at the forward edge
of the wedge show that consideration of physical and chemical
transformations also yields a sharp reduction in the angle of
the attached wave and increases the speed at which the detached
wave becomes an attached wave on increasing the wave velocity.

In an ideal gas with $\gamma = 1.4$ no attached shocks
exist for $\alpha_o > 18°$. For version 1, which takes into account
all the molecular transformations possible at the given temperature, the limiting angle for detachment of the shock from the
nose of the wedge body is 42°. This means that the change in the
ratio of specific heats brought about by chemical and physical
transformations changes appreciably the gas-dynamic process attendant to flow of hot gas about a wedge. Thus, when an ideal
gas heated by a shock wave flows past the wedge at $u_o = 1900$ m/
sec (version 2), a detached wave is formed ; conversely, an
attached wave is formed in a gas with vibrational molecular
levels excited.

The variation in ω_2 as a function of the extent
of physical and chemical transformations is more pronounced
than that in the angle of attached wave. In this case conditions

no longer exist for smooth transition from the reflected to the attached wave, a break forms and consequently, a secondary Mach wave is formed.

In the case of $\omega_2 > 0$, it is geometrically possible to have a smooth joining of the attached and reflected waves. However, as is shown experimentally, this kind of transition does not always occur and a break does form in the reflected wave. The point at which the secondary break forms can be estimated as follows. In the coordinate system moving with the triple point the gas velocity behind the reflected wave is always supersonic. Let us find the boundary of the effect of signals which disturb the region behind the reflected wave. Along the tangential discontinuity line we lay off velocity u_0 and find the point of intersection of the reflected wave with a circle of radius a_2 . We determine the difference $x_1 - x_2$. It may be assumed that the break on the reflected wave which is due to the need for joining the processes of reflection and flow past the vertex of the wedge will take place in point R which defines the region of influence of angular signals in the gas flow behind the reflected wave. The region of influence was calculated for version 2 in air. The theoretically calculated data were compared with the experimentally obtained values of $x_1 - x_2$. It can be seen that the experimental points lie on the theoretical curve. Knowing the position of the second triple point it is possible to determine from the three-shock theory

the direction and strength of the secondary Mach wave in the
vicinity of the second triple point.

The variation in the ratio of specific heats γ
in a gas heated by a shock wave results in appreciable changes
not only of the Mach configuration proper. It also affects the
transition from regular to Mach reflections.

Calculations show that excitation of internal
degrees of freedom in the gases under study displaces the limiting angle of transition of regular into Mach reflection by 1-3°,
increasing the range of regular reflection.

6. MEASUREMENT

The measurements were carried out in a 72 mm
square shock tube. A 24° angle wedge was mounted in the shock
tube. The shock velocity changed from 600 to 2500 m/sec, the
incident pressure being 12 mm H_g. A 1 mm pressure gauge and a
thin film platinum gauge were located in the wedge surface.
Experimental methods were described in [5]. The experiments were
made in air and in carbon dioxyde.

It has been found that the pressure distribution
was not what would be expected from low speed Mach reflection
experiments. Ther is an increase in pressure downstream from
the Mach stem. There are two steps in oscilloscope traces ;
(Fig. 3) The pressure rise associated with the first step is

due to the pressure increase in the Mach wave. An additional
pressure rise has been observed only for strong wave Mach reflection. The same phenomenon was observed in [5, 8, 9]. The
pressure ratio of the second pressure step to the first step
rises with the incident shock velocity (Fig. 4). In air, an
additional pressure rise can be seen at the incident velocity
greater than 900 m/sec.

In Fig. 5 the results of surface temperature
measurements are shown for the double Mach reflection. The
change of the surface temperature with time is due to the change
of the heat flux from the hot gas to the glass. The heat flux
into the metal of the surface can be deduced from the measurements of the heat flux to the glass. The temperature was measured with the aid of a thin film gauge with 6 per cent accuracy.
Keller and Ryan's method [10] was used for calculating the heat
flux. The experimental temperature – time curve was approximated
with a number of segments. The slope of the segments was used in
the heat flux equation.

Heat flux distribution along the wedge surface
can be seen in Fig. 6. After the first rise due to the Mach wave
there is an additional peak of heat flux at the moment of the
secondary temperature step. As the incident shock velocity rises,
the additional peak becomes greater.

Fig. 7 shows the time between the two steps as
measured by the temperature gauges (close points) and the pres-

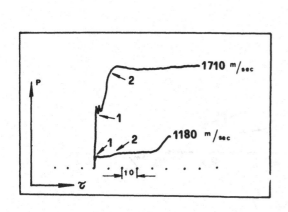

Fig. 3

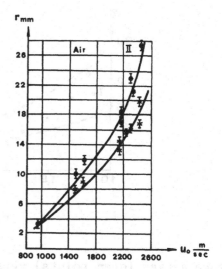

Fig. 4

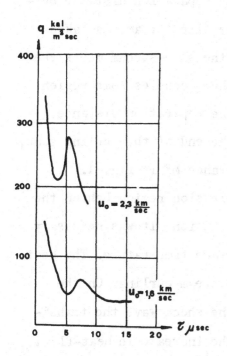

Fig. 5

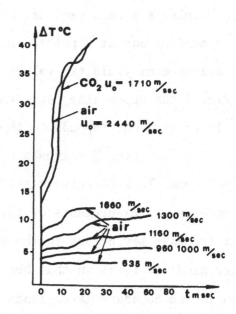

Fig. 6

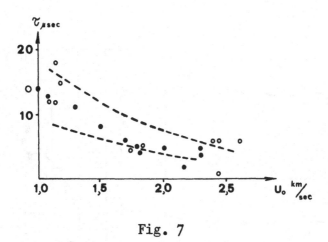

Fig. 7

sure gauges (open points) versus the incident wave velocity. The
time was observed to correspond to the separation distance between the foot of the Mach stem and the slip-stream, as taken
from Toepler's streak-records. Since the slip-stream meets the
wedge surface not at a point, but rather occupies some region,
the measurement yield two values of the separation distance.
In fig. 7 the upper line represents the end of the curling zone,
the lower line corresponds to the distance ME in Fig. 1.

          Fig. 7 suggests a compression region behind the
slip-stream. To substantiate this conclusion, it was necessary
to make extra measurements of the surface temperature. The heat
flux increase may be due to the slip-stream curling. On the
other hand, it is known that behind the shock wave the transition in the boundary layer leads to the increase in heat-flux.

          Reynolds number behind Mach wave was calculated
for the experiments. In conditions under study Reynolds number

was found to change from 10^4 to $5 \cdot 10^4$ at the distance from Mach

stem foot where the additional temperature rise was observed.

These values are lower than the transition one, as taken from

literature. Therefore, at the initial pressure of $12.7\,mm\,Hg$ the

boundary layer behind Mach wave is laminar up to the contact

surface and the secondary heat-flux rise is due to the existen-

ce of the hot and compressed region behind the curling slip-

stream. A further proof was obtained with experiments at great-

er initial pressure ($100\,mm\,Hg$). In this case Reynolds number

can reach its transition value and the transition from laminar

to turbulent boundary layer can occur in the region between

Mach wave and the slip-stream. That was actually the case. The

oscilloscope temperature traces showed two additional steps, the

second occuring at the time of the slip-stream passing the gauge

bare. The transition Reynolds number which was obtained from

these experiments was in good agreement with the results of

other authors and was an order of magnitude greater than the

Reynolds numbers at the time of the additional step appearance

in the experiments with the initial pressure being $12.7\,mm\,Hg$.

7. INTERPRETATION OF EXPERIMENTAL OBSERVATIONS

The temperature and pressure measurements show

that there is an additional pressure region behind the foot of

the slip-stream. It is obvious that for double Mach reflection

with the second shock wave (Fig. 1 a) the pressure and temperature in the region 6 behind the secondary shock wave RS is expected to rise. But the additional steps in pressure and tempera
ture traces were observed to exist not only in the case (Fig. 1
a), but also in the case of Mach pattern without the second
triple point (Fig. 1 b). In air, the second pressure increase
was observed at incident shock velocities greater than 900 m/sec
and the second temperature rise at the velocity of about 630 m/sec

At the same time the transition from the pattern (a) to
the pattern (b) was observed to occur at the velocities about
1400 m/sec which means that even in absence of secondary shock
wave, the compression region behind the shock was developed.

The interpretation of the mechanism that accounts
for the above additional compression and for the secondary shock
wave is to be sought in the peculiarities of the Mach reflection
mentioned by White and Courant [9,11].

From simple physical considerations, the gas velocities in region 2 and 6' are expected to be different from
each other. One can calculate the gas velocity in regions 2 and
3 from the three-shock theory. The calculations show that the
gas is in rest in region 3 relative to the point E – the point
of intersection of the slip-stream with the wedge surface (Fig.
8), but the gas velocity in region 2 is not equal to zero and
the gas moves along the slip-stream. But the gas at the wedge
surface must move along the surface with the speed equal to the

speed of the contact surface, that is, the speed of the gas
flow in the system of reference bound with the point E must
be zero.

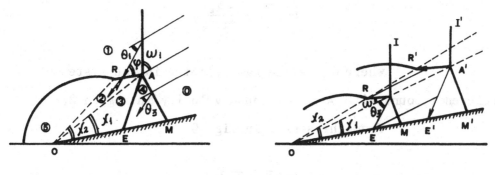

Fig. 8   Fig. 9

Therefore the gas moving from region 2 into region 6' must
decelerate relative to the point E . The pressure increase in
this process can be estimated and in order to carry out the estimation the system of reference bound with the point E must
be used.

Fig. 8 shows the pattern and angle indications
in Mach configuration. In order to go over to the system of
reference bound with E point, it is necessary to calculate E
point velocity in the first triple point system of reference.
Fig. 9 shows two successive positions of the configuration. It
can be seen that the speed is proportional to the segment AE
and is directed along the slip-stream.

From the triangle OAE

$$a_E = \frac{u_o}{\sin \omega_1} \frac{\sin \chi}{\sin(\chi + \theta)}$$

So the Mach number in region 2 in the reference system bound with E is equal to

$$M_\varepsilon = \frac{\bar{u}_2 - a_\varepsilon}{a_2} \, ,$$

where $\bar{u}_2$ – the gas velocity in the system of reference bound with A, a_2 –sound velocity in region 2.

From the scheme in Fig. 9

$$\alpha_\varepsilon = x_1 + \theta_3 .$$

Velocities u_1 and u_2 are found from the usual conservation laws.

The stagnation pressure in the system bound with the point E is

$$p_{6'} = p_2 + \rho \, (\bar{u}_2 - a_\varepsilon)^2 .$$

Therefore, for experimentally known values of u_0 and x_1 one can estimate the pressure $p_{6'}$ using the three-shock theory and taking into account real gas properties [5] .

Following this procedure the stagnation pressure in the region $6'$ was estimated for different velocities of incident waves.

The calculated pressure ratio $p_{6'}/p_3$ in air is shown in Fig. 10 versus the incident shock velocity. Experimentally found pressure ratios are in fair agreement with the calculated values.

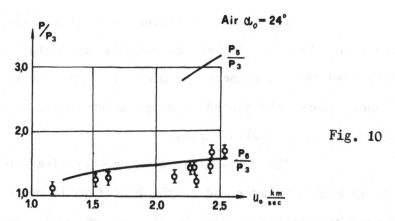

Fig. 10

The pressure behind the slip stream must grow after the increase in region 6′ along the surface towards the wedge apex for at high velocities the pressure behind bow shock wave P_5 must be greater than the pressure behind Mach wave P_3. The calculated value P_5/P_3 is shown in Fig. 10 for these velocities of incident shock waves where the bow waves are already attached.

As already mentioned above, in air at the incident shock velocity near 1400 m/sec, there occurs a transition to the flow pattern with the secondary shock wave. Therefore, in order to determine the secondary pressure increase it is necessary to take account of the pressure increase due to this process as well. But in air the secondary shock is weak and the pressure increase is insignificant. This accounts for good agreement between the calculated curve and the experimental data in Fig. 10. In CO_2 on the other hand the account of the pressure increase in the secondary shock wave is indispensable.

In stagnation region 6' , in addition to pressure rise, there is also an increase in temperature and density, which results in the increased heat flux onto the wedge surface. As shown above, the secondary compression region develops just behind the curling slip-stream.

The slip-stream curling would lead to the transition in boundary layer and to the heat-flux increase due to increased heat transfer even though the compression region did not exist at all. When estimating the lower value of heat transfer coefficient may be assumed to be constant. Then ratio of the heat flux into the wall from region 6' to the heat flux from region 3 is equal to

$$\frac{q_{6'}}{q_3} = \frac{\rho_{6'} u_{6'}}{\rho_3 u_3} \frac{(h - h_w)_{6'}}{(h - h_w)_3}$$

Here ρ –density, u – gas velocity, h –stagnation enthalpy, h_w – gas flow enthalpy at the wedge surface. When thus calculated, the heat flux ratios were shown to range from 1.3 to 1.6 at the velocity changing from 1.0 to 2.5 km/sec. When we calculated from the experimental curves q vst (Fig. 6) the ratio of the second maximum to the preceeding minimum, all the experimental data were found to range from 1.3 to 3.0 in the same velocity range. That is, the calculation did yield the lower value of the heat transfer ratio. In CO_2 in presence of the second shock wave the heat flux increase must be still greater due to gas heating in the shock.

REFERENCES

[1] Gvozdeva L.G., Bazhenova T.V., Predvoditeleva O.A. and
Fokeev V.P. Mach Reflection of Shock Waves in Real Gases, Astronautica Acta, v. 14, p. 503–508 (1969).

[2] Gvozdeva L.G. and Predvoditeleva O.A. An Experimental Study of the Mach Reflection of Shock Waves at Velocities 1000–3000 m/sec, in Carbon Dioxide, Nitrogen and Air, Dokl. Akad. Nauk USSR 163, 1088 (1965) Soviet Phys. Dokl. 10, 694 (1966)

[3] Gvozdeva L.G., and Predvoditeleva O.A. Mach Reflection of Shock Waves Moving with the Speed about 2000 m/sec in Carbon Dioxide and Nitrogen. In Research into Physical Gas Dynamics p. 183 Nauka, Moscow (1966).

[4] Gvozdeva L.G., Predvoditeleva O.A. and Fokeev V.P. Double Mach Reflection of Strong Shock Waves, Izv. Akad. Nauk USSR, Mech. zhidk. i gas. p. 1 p. 90–112 (1968).

[5] Bazhenova T.V., Gvozdeva L.G., Lobastov Yu.S., Naboko I.M., Nemkov R.G. and Predvoditeleva O.A. Udarnye vonly v realnykh gazakh. p. 73 Moscow, Nauka Press (1968) Trans Shock Waves in Real Gases, NASA TTF-585, Washington, D.C. 20546, October 1969.

[6] Weynants R.R. An Experimental Investigation of Shock Wave Diffraction over Compression and Expansion Corners, UTIAS Technical Note, April 1968.

[7] Sytchikova M.P., Berezkina M.K. and Semenov A.N.
     Flow Formation around the Model in Shock Tube,
     Aerophys. issledovaniya sverchzvukovych techenii.
     p. 7-13. Nauka Press, M.L. (1967).

[8] White D.R. An Experimental Survey of the Mach Reflection
     of Shock Waves, Proc. II Midwest Conf. Fluid.
     Mech. v. XVI, p. 253 (1952).

[9] Merritt D.L. Mach Reflection on a Cone, AIAA Journal
     v. 8 No. 6, p. 1208-1209 (1968).

[10] Keller J.A. and Ryan N.W. ARS Journal v. 31, No. 10
      (1961).

[11] Courant R. and Freidrichs K.O. Supersonic Flow and Shock
      Waves, Pure and Applied Mathematics v. 1, p. 318-
      350 Interscience, New York (1948).

CONTENTS

Printed in the United States
By Bookmasters